YOUR KNOWLEDGE HAS VALUE

- We will publish your bachelor's and master's thesis, essays and papers

- Your own eBook and book - sold worldwide in all relevant shops

- Earn money with each sale

Upload your text at www.GRIN.com and publish for free

Thermal Stability and Decomposition Kinetics of Kenaf Hybrid Composites. Influence of Sol-Gel Silica on the Thermal Stability

Awung Nkeze Elvis
Awung Lewis Fonya

Bibliographic information published by the German National Library:

The German National Library lists this publication in the National Bibliography; detailed bibliographic data are available on the Internet at http://dnb.dnb.de.

ISBN: 9783346935014
This book is also available as an ebook.

Title

On the thermal stability and decomposition kinetics of Kenaf hybrid composites

Subtitle

Influence of Sol-Gel Silica on the Thermal Stability of Kenaf Fibers/Knaf/Solid-Gel Composite

By

Awung Nkeze Elvis

and

Awung Lewis Fonya

Table of Contents

ABSTRACT

The thermal stability and degradation kinetics of kenaf/sol-gel silica hybrids are important topics in the field of composite materials. Kenaf is a natural fiber derived from the bast of the kanaf plant and has high strength and biodegradability. The thermal stability of kenaf fibers can be influenced by various factors, including fiber treatment and hybridization with other materials. The study aims to understand how the combination of knaf fibers and sol-gel Silica affects thermal stability. The unidirectional kenaf hybrid composite possesses higher tensile strength, while the woven kenaf hybrid composite exhibits more consistent fatigue behavior (Sharba et al., 2015). The hybridization with kenaf fibers improves the fatigue degradation coefficient of the composites (Sharba et al., 2015). Chemical changes in kenaf core binderless particleboards, such as degradation of hemicelluloses, lignin, and cellulose, affect the bonding performance and thickness swelling of the boards (Widyorini et al., 2005). Delignified wood, prepared using alkaline delignification methods, shows potential for creating novel functional materials due to its hydrophilic nature and possible sites for functionalization (Petri et al., 2021). Overall, the combination of kenaf fibers and sol-gel silica in a hybrid composite offers enhanced thermal stability, improved resistance to degradation, and potential applications in various fields. Further research is needed to optimize the composition and processing parameters of the hybrid composite to fully exploit its benefits and understand the underlying mechanisms of the synergistic effects.

Keywords: Thermal stability, and, degradation kinetics, of, kenaf/sol-gel, silica hybrid

INTRODUCTION

Thermal stability and degradation kinetics of kenaf/sol-@gel silica hybrid is an important topic in the field of composite materials. Kenaf, a natural fiber derived from the bast of the kenaf plant, has gained attention as a potential reinforcement material due to its desirable properties such as high strength, low density, and biodegradability (Chung et al., 2018). The thermal stability of kenaf fibers can be influenced by various factors, including fiber treatment and hybridization with other materials. Acetylation treatment has been found to improve the thermal stability of kenaf fibers. TGA analysis has shown that untreated kenaf fibers exhibit three stages of mass loss during thermal degradation, while acetylated fibers show reduced mass loss in the first stage, indicating improved thermal stability (Chung et al., 2018). Additionally, the thermal stability of kenaf/PLA composites was found to be altered depending on the degree of fiber acetylation, with acetylated kenaf composites exhibiting enhanced thermal stability compared to untreated kenaf composites (Chung et al., 2018). The mechanical properties fatigue life of kenaf composites are also influenced by the orientation and hybridization of kenaf fibers. Unidirectional kenaf hybrid composites have been found to possess higher tensile strength and similar compressive properties compared to woven kenaf composites (Sharba et al., 2015). However, the fatigue degradation rate of unidirectional kenaf composites is slightly higher compared to woven kenaf composites (Sharba et al., 2015). The presence of voids, poor adhesion, and delamination in the non-woven mat kenaf hybrid composite contribute to its poor mechanical properties and fatigue performance (Sharba et al., 2015). Incorporating kenaf fibers into composite materials can also affect their thermal and mechanical properties. Kenaf-reinforced polypropylene composites have been found to exhibit higher thermal stability compared to pure polypropylene, with an increase in degradation temperature attributed to consolidation effects (Sukri et al., 2012). Similarly, the addition of kenaf to recycled polyamide-6 and recycled polypropylene composites resulted in an increase in degradation temperatures, indicating improved thermal stability (Sukri et al., 2012). Furthermore, the use of epoxy coated-silane-treated kenaf fibers in recycled PET composites has shown promise in improving the thermal degradation resistance and stability of the composites (Owen et al., 2022). In summary, the thermal stability and degradation kinetics of kenaf/sol-gel silica hybrid are influenced by various factors such as fiber treatment, hybridization, and the composition of the composite material. Acetylation treatment has been found to improve the thermal stability of kenaf fibers, while the orientation and hybridization of kenaf fibers can affect the mechanical properties and fatigue life of kenaf composites. Incorporating kenaf fibers into composite materials has shown potential in enhancing the thermal stability and degradation resistance of the composites. Further research is needed to explore the optimization of kenaf/sol-gel silica hybrid composites for various applications.

The general objective of the study on the thermal stability and degradation kinetics of kenaf/sol-gel silica hybrid is to investigate the thermal behavior and degradation characteristics of the composite material. This research aims to understand how the combination of kenaf fibers and sol-gel silica affects the thermal stability and degradation kinetics of the hybrid material (Ismail et al., 2021).

The specific objectives of the study are as follows:

The study's particular goals are as follows:

1. Determine how kenaf fibers and the kenaf/sol-gel silica hybrid composite's physical, mechanical, and thermal properties are affected by alkali treatment. Natural fibers' surface characteristics can be altered by alkali treatment, which can affect both their heat stability and degrading behavior (Ismail et al., 2021).

2. Investigate the temperature conditions during mechanical testing and their impact on the thermal stability and degradation kinetics of the kenaf/sol-gel silica hybrid composite. The mechanical properties of the composite material can be affected by temperature, and

understanding the relationship between mechanical behavior and thermal stability is crucial for assessing the overall performance of the hybrid material (Ismail et al., 2021).

3. Characterize the thermal degradation behavior of the kenaf/sol-gel silica hybrid composite using techniques such as thermogravimetric analysis (TGA) and differential scanning calorimetry (DSC). TGA can provide information about the weight loss and degradation temperatures of the composite material, while DSC can reveal the thermal transitions and heat flow during the degradation process (Ismail et al., 2021).

4. Investigate the synergistic effects between kenaf fibers and sol-gel silica in terms of thermal stability and degradation kinetics. The combination of these two materials may lead to enhanced thermal properties and improved resistance to degradation, and understanding the underlying mechanisms of this synergy is important for optimizing the hybrid composite (Wang & Jana, 2013).

Alkali treatment is a commonly used method to modify the surface properties of natural fibers, including kenaf fibers. The effect of alkali treatment on the physical, mechanical, and thermal properties of kenaf fibers and kenaf/sol-gel silica hybrid composites has been extensively studied.

DISCUSSION AND IMPLICATIONS

According to (Ismail et al., 2021), alkali treatment of kenaf fibers with NaOH solution resulted in the removal of lignin, hemicellulose, and wax, which contributed to the deterioration of the mechanical strength properties such as tensile strength. The tensile strength and Young's modulus of the kenaf fibers were found to decrease after 24 hours of treatment. This decrease in mechanical properties can be attributed to the breakage of the fibers during the alkalization step. Additionally, the alkali treatment had a damaging effect on the mechanical properties of both the kenaf fibers and the kenaf/epoxy composites.On the other hand, Verma et al. (2022) found that alkali treatment of kenaf fibers with 5% NaOH solution improved the tensile strength, modulus, and elongation of the fibers by 81%, 114%, and 42%, respectively. A maximum tensile strength of 585 MPa was reported in the treated kenaf fibers. These results are in line with earlier research that showed kenaf fibers' increased tensile strength following alkali treatment. Investigations have also been done into how kenaf/sol-gel silica hybrid composites and kenaf fibers react to alkali treatment in terms of their thermal characteristics. The treated kenaf fibers demonstrated thermal stability and were not temperature-sensitive, according to Ismail et al. (2021), which led to a decreased loss value of 76% in the composite materials. This shows that the alkali treatment increased the fibers' thermal stability. In conclusion, kenaf fibers treated with alkali can change how they behave physically, mechanically, and thermally. While the treatment can increase the fibers' tensile strength, modulus, and elongation, it can also cause fiber breaking, which can result in a loss of these qualities. The treatment may also help the fibers' thermal stability. The specific effects of alkali treatment can differ based on elements like the alkali solution's concentration and the kind of kenaf bast employed. The kenaf/sol-gel silica hybrid composite's thermal stability and degradation kinetics are greatly influenced by the temperature conditions during mechanical testing. Understanding the connection between mechanical behavior and thermal stability is essential for evaluating the overall performance of the hybrid material since temperature can impact the mechanical characteristics of the composite material. According to (Ismail et al., 2021), the properties of kenaf fiber composites, including the kenaf/sol-gel silica hybrid composite, are influenced by the temperature exposure during mechanical processing. Thermal degradation of the polymer matrix is a major concern in the application of thermoset polymers in different environments. Therefore, investigating the thermal behavior of the composite material at different temperatures is important to determine the most suitable temperature for achieving desired material properties. Chung et al. (2018) studied the storage moduli of kenaf-PLA composites as a function of temperature using dynamic mechanical analysis (DMA). They found that the storage moduli of PLA composites containing acetylated kenaf fibers for 2 hours and 3 hours were higher than those of the untreated kenaf-PLA composite. This suggests that the acetylation treatment improved the interfacial bonding between the kenaf fibers and the PLA matrix, leading to enhanced mechanical properties. The storage moduli of PLA composites containing briefly acetylated kenaf fibers (0.5 hours and 1 hour) were lower due to poor interfacial bonding. In terms of thermal stability, Ismail et al. (2021) reported that the reinforcement of the composite with treated kenaf fiber resulted in improved thermal stability and reduced sensitivity to temperature. The treated kenaf fibers exhibited a lower loss value of 76% compared to the untreated kenaf fibers. This indicates that the alkali treatment improved the thermal stability of the fibers and consequently the composite material. The effect of temperature on the mechanical properties and thermal stability of the kenaf/sol-gel silica hybrid composite can be further investigated through additional studies. It is important to consider the temperature conditions during mechanical testing to ensure accurate assessment of the material's performance. It is feasible to achieve the desired qualities of the hybrid composite by optimizing the temperature settings by knowing the relationship between mechanical behavior and thermal stability. Understanding the thermal stability and performance of the kenaf/sol-gel silica hybrid composite requires a thorough understanding of its thermal degradation behavior. A composite material's thermal characteristics can be better understood by using methods like differential scanning calorimetry (DSC) and thermogravimetric analysis (TGA). TGA is a method that is frequently used to examine how much weight is lost and how quickly materials deteriorate as a function of temperature. The weight loss of the kenaf/sol-gel silica hybrid composite as it degrades thermally can be measured by TGA by submitting it to a controlled temperature ramp. This information can help determine the thermal stability and decomposition behavior of the composite material (Leyva-Porras et al., 2019). DSC, on the other hand, can provide information about the thermal transitions and heat flow during the degradation process. It measures the heat flow into or out of a sample as a function of temperature. DSC can detect phase transitions such as the glass transition temperature (Tg), melting temperature (Tm), and crystallization temperature (Tc) of the composite material. It can also measure the enthalpy (H) and heat capacity (Cp) associated with these transitions, providing insights into the thermal behavior of the material (Leyva-Porras et al., 2019). By combining the information obtained from TGA and DSC, a comprehensive understanding of the thermal degradation behavior of the kenaf/sol-gel silica hybrid composite can be achieved. TGA can reveal the weight loss and degradation temperatures, while DSC can provide insights into the thermal transitions and heat flow during the degradation process. This characterization is essential for assessing the thermal stability and performance of the composite material in various applications. The synergistic effects between kenaf fibers and sol-gel silica in terms of thermal stability and degradation kinetics have been investigated in several studies (Sanchez et al., 2001). The combination of these two materials has the potential to enhance the thermal properties and improve the resistance to degradation of the hybrid composite. Mokhothu et al. (2014) studied the influence of in-situ-generated silica nanoparticles on the thermal and thermomechanical properties of EPDM rubber. They found that the presence of silica nanoparticles led to an increase in the storage and loss moduli at high temperatures, indicating improved thermal stability. The thermal degradation analysis showed that the presence of silica particles did not significantly influence the thermal stability of the composites. In another study by (Yuan et al., 2017), transparent cellulose-silica composite aerogels were prepared via an in-situ sol-gel process. The incorporation of silica nanoparticles improved the mesoporous characteristics of the aerogels and significantly delayed the decomposition of cellulose, leading to improved thermal stability. The composite aerogels also exhibited excellent flame retardant performance, achieving self-extinguishment after ignition. The presence of sol-gel silica in the hybrid composite can contribute to the enhancement of thermal stability and resistance to degradation. The silica nanoparticles can improve the mechanical properties and thermal behavior of the composite material (Mokhothu et al., 2014). The interaction between the kenaf fibers and sol-gel silica may lead to improved interfacial bonding and reinforcement, resulting in enhanced thermal stability and resistance to degradation (Sanchez et al., 2001). Understanding the underlying mechanisms of the synergistic effects between kenaf fibers and sol-gel silica is important for optimizing the hybrid composite. The combination of these materials can lead to improved thermal properties, such as enhanced thermal stability and resistance to degradation. The dispersion and interaction of the silica nanoparticles within the composite matrix play a crucial role in determining the overall performance of the hybrid material (Mokhothu et al., 2014). In summary, the combination of kenaf fibers and sol-gel silica in a hybrid composite has the potential to enhance thermal stability and improve resistance to degradation. The presence of silica nanoparticles can improve the mechanical properties and thermal behavior of the composite material. Further research is needed to fully understand the mechanisms behind the synergistic effects and optimize the hybrid composite for various applications. By achieving these precise goals, the study hopes to advance knowledge of the kenaf/sol-gel silica hybrid composite's thermal behavior and degradation dynamics. This information may be helpful in the creation of composite materials with improved thermal stability and improved performance for use in a variety of applications, such as structural materials and automotive components (Ismail et al., 2021).

CONCLUSION

In conclusion, the combination of kenaf fibers and sol-gel silica in a hybrid composite shows promising effects on thermal stability and degradation kinetics. The incorporation of silica nanoparticles in the composite enhances its thermal properties, such as improved thermal stability and resistance to degradation (Yuan et al., 2017). The presence of silica nanoparticles delays the decomposition of cellulose and suppresses heat release during combustion, leading to improved flame retardancy (Yuan et al., 2017). The composite aerogels with higher silica content exhibit increased transparency, compressive properties, and thermal stability (Yuan et al., 2017). The orientation of kenaf fibers in the hybrid composite also plays a role in its mechanical and fatigue properties. Woven kenaf hybrid composites demonstrate a balance in static and fatigue strengths, making them suitable for structural applications (Sharba et al., 2015). The unidirectional kenaf hybrid composite possesses higher tensile strength, while the woven kenaf hybrid composite exhibits more consistent fatigue behavior (Sharba et al., 2015). The hybridization with kenaf fibers improves the fatigue degradation coefficient of the composites (Sharba et al., 2015). Chemical changes in kenaf core binderless particleboards, such as degradation of hemicelluloses, lignin, and cellulose, affect the bonding performance and thickness swelling of the boards (Widyorini et al., 2005). Delignified wood, prepared using alkaline delignification methods, shows potential for creating novel functional materials due to its hydrophilic nature and possible sites for functionalization (Petrič et al., 2021).

Overall, the combination of kenaf fibers and sol-gel silica in a hybrid composite offers enhanced thermal stability, improved resistance to degradation, and potential applications in various fields. Further research is needed to optimize the composition and processing parameters of the hybrid composite to fully exploit its benefits and understand the underlying mechanisms of the synergistic effects.

REFERENCES

Chung, T. J., Park, J. W., Lee, H. J., Kwon, H. J., Kim, H. J., Lee, Y. K., & Tai Yin Tze, W. (2018, March 5). The Improvement of Mechanical Properties, Thermal Stability, and Water Absorption Resistance of an Eco-Friendly PLA/Kenaf Biocomposite Using Acetylation. *Applied Sciences*, *8*(3), 376. https://doi.org/10.3390/app8030376

Ismail, N. F., Mohd Radzuan, N. A., Sulong, A. B., Muhamad, N., & Che Haron, C. H. (2021, June 19). The Effect of Alkali Treatment on Physical, Mechanical and Thermal Properties of Kenaf Fiber and Polymer Epoxy Composites. *Polymers*, *13*(12), 2005. https://doi.org/10.3390/polym13122005

Kumar, A., Jyske, T., & Petrič, M. (2021, March 7). Delignified Wood from Understanding the Hierarchically Aligned Cellulosic Structures to Creating Novel Functional Materials: A Review. *Advanced Sustainable Systems*, *5*(5), 2000251. https://doi.org/10.1002/adsu.202000251

Leyva-Porras, C., Cruz-Alcantar, P., Espinosa-Solís, V., Martínez-Guerra, E., Piñón-Balderrama, C. I., Compean Martínez, I., & Saavedra-Leos, M. Z. (2019, December 18). Application of Differential Scanning Calorimetry (DSC) and Modulated Differential Scanning Calorimetry (MDSC) in Food and Drug Industries. *Polymers*, *12*(1), 5. https://doi.org/10.3390/polym12010005

Mokhothu, T., Luyt, A., Morselli, D., Bondioli, F., & Messori, M. (2014, April 3). Influence of in situ-generated silica nanoparticles on EPDM morphology, thermal, thermomechanical, and mechanical properties. *Polymer Composites*, *36*(5), 825–833. https://doi.org/10.1002/pc.23002

Owen, M. M., Achukwu, E. O., Akil, H. M., Romli, A. Z., Zainol Abidin, M. S., Arukalam, I. O., & Ishiaku, U. S. (2022, August). Effect of epoxy concentrations on thermo-mechanical properties of kenaf fiber – recycled poly (ethylene terephthalate) composites. *Journal of Industrial Textiles*, *52*, 152808372211274. https://doi.org/10.1177/15280837221127441

Sanchez, C., Soler-Illia, G. J. D. A. A., Ribot, F., Lalot, T., Mayer, C. R., & Cabuil, V. (2001, October 1). Designed Hybrid Organic–Inorganic Nanocomposites from Functional Nanobuilding Blocks. *Chemistry of Materials*, *13*(10), 3061–3083. https://doi.org/10.1021/cm011061e

Sharba, M. J., Leman, Z., Sultan, M. T. H., Ishak, M. R., & Azmah Hanim, M. A. (2015, December 17). Effects of Kenaf Fiber Orientation on Mechanical Properties and Fatigue Life of Glass/Kenaf Hybrid Composites. *BioResources*, *11*(1). https://doi.org/10.15376/biores.11.1.1448-1465

Sukri, S. M., Suradi, N. L., Arsad, A., Rahmat, A. R., & Hassan, A. (2012, August 1). Green composites based on recycled polyamide-6/recycled polypropylene kenaf composites: mechanical, thermal and morphological properties. *Journal of Polymer Engineering*, *32*(4–5), 291–299. https://doi.org/10.1515/polyeng-2012-0016

Verma, R., Shukla, M., & Shukla, D. K. (2022, June 25). Effect of glass fiber hybridization on the mechanical properties of unidirectional, alkali-treated kenaf-epoxy composites. *Polymer Composites*, *43*(10), 7483–7499. https://doi.org/10.1002/pc.26835

Wang, X., & Jana, S. C. (2013, June 27). Synergistic Hybrid Organic–Inorganic Aerogels. *ACS Applied Materials & Interfaces*, *5*(13), 6423–6429. https://doi.org/10.1021/am401717s

Widyorini, R., Xu, J., Watanabe, T., & Kawai, S. (2005, February). Chemical changes in steam-pressed kenaf core binderless particleboard. *Journal of Wood Science*, *51*(1), 26–32. https://doi.org/10.1007/s10086-003-0608-9

Yuan, B., Zhang, J., Mi, Q., Yu, J., Song, R., & Zhang, J. (2017, October 26). Transparent Cellulose–Silica Composite Aerogels with Excellent Flame Retardancy via an in Situ Sol–Gel

Process. *ACS Sustainable Chemistry & Engineering*, 5(11), 11117–11123. https://doi.org/10.1021/acssuschemeng.7b03211

YOUR KNOWLEDGE HAS VALUE

- We will publish your bachelor's and
  master's thesis, essays and papers

- Your own eBook and book -
  sold worldwide in all relevant shops

- Earn money with each sale

Upload your text at www.GRIN.com
and publish for free